T0269445

CAMBRIDGE LIBRARY COLLECTION

Books of enduring scholarly value

Botany and Horticulture

Until the nineteenth century, the investigation of natural phenomena, plants and animals was considered either the preserve of elite scholars or a pastime for the leisured upper classes. As increasing academic rigour and systematisation was brought to the study of 'natural history', its subdisciplines were adopted into university curricula, and learned societies (such as the Royal Horticultural Society, founded in 1804) were established to support research in these areas. A related development was strong enthusiasm for exotic garden plants, which resulted in plant collecting expeditions to every corner of the globe, sometimes with tragic consequences. This series includes accounts of some of those expeditions, detailed reference works on the flora of different regions, and practical advice for amateur and professional gardeners.

Cocoa

In Britain, the name of Cadbury has been synonymous with chocolate ever since John Cadbury opened his factory in 1831. This book, written by Richard Cadbury (1835–99) under the pen name 'Historicus', was published in 1892. It describes the natural history of the tropical American cocoa plant, its spread in cultivation across the world, and the history of its use, before dealing with the manufacturing process, as exemplified by the Cadbury factory at Bournville, surrounded by the model housing and leisure facilities which the family built for its workers. The processing of cocoa beans into solid and drinking chocolate is described in detail, with emphasis on the developments in machinery which simplified production. A chapter deals with the importance of the vanilla plant for flavouring, and an appendix gives guidance on the cultivation of cocoa trees. This remains a fascinating account of one of the world's most popular indulgences.

Cocoa

All About It

RICHARD CADBURY

CAMBRIDGE
UNIVERSITY PRESS

University Printing House, Cambridge, CB2 8BS, United Kingdom

Cambridge University Press is part of the University of Cambridge.
It furthers the University's mission by disseminating knowledge in the pursuit of education, learning and research at the highest international levels of excellence.

www.cambridge.org
Information on this title: www.cambridge.org/9781108082273

This edition first published 1892
This digitally printed version 2015

ISBN 978-1-108-08227-3 Paperback

The original edition of this book contains a number of colour plates, which have been reproduced in black and white. Colour versions of these images can be found online at www.cambridge.org/9781108082273

Reproduction of the Frontispiece of a very old Latin Book on Chocolate, published in 1639. The subject is allegorical, representing a Casket of Chocolate being handed to Neptune to make known to the Countries of the World.

The material originally positioned here is too large for reproduction in this reissue. A PDF can be downloaded from the web address given on page iv of this book, by clicking on 'Resources Available'.

COCOA:

ALL ABOUT IT.

BY "HISTORICUS."

LONDON:
SAMPSON LOW, MARSTON & COMPANY,
Limited,
ST. DUNSTAN'S HOUSE,
FETTER LANE, FLEET STREET, E.C.
1892.

CONTENTS.

LIST OF ILLUSTRATIONS.

COCOA:

ALL ABOUT IT.

—:o:—

CHAPTER I.

—:o:—

HISTORY AND CULTIVATION OF THE PLANT.

—:o:—

IT was one of the dreams of our childhood to sail on the bosom of that mighty river whose water-shed covers the greater part of the northern portion of the continent of South America, and to explore into the secrets of its thousand tributaries that penetrate into forests untrodden by the foot of man, teeming with innumerable brilliantly-coloured birds and insects, luxuriating in their own Paradise of tropical plants and flowers. Far into the dark recesses of these forests the tributary streams of the Amazon flow, shadowed by forest trees growing to the water's edge, festooned by gigantic creepers which hang in rich foliage and flower over them.

We follow them further on to their sources among the snow fields and rocky defiles of the Andes, and amidst the ruins of an ancient world and people, almost extinct as nations, but whose history brings back thrilling stories of bye-gone days of civilization and government.

This was the original home of the Cocoa plant, and it is found at the present day in its wild state both on the banks of the Amazon, in Mexico, and in the United States of Columbia.

Cacao Flowers and Pods, showing inside of Pod.—(*Drawn from Nature*).

Its growth is now distributed over a great portion of the tropical world, and it will thrive within 25 parallels of latitude, but

Leaves, Flowers, and Fruit.

luxuriates within 15, and is cultivated as high as 1,700 feet above the level of the sea.

The largest quantities of Cocoa are produced in Guayaquil, Para, and Bahia, the West Indies, Ceÿlon, and some portions of the continent of Africa.

The finest qualities are grown in Central America, Trinidad, and Ceylon; the latter is of comparatively recent cultivation, but is the most delicate in colour, flavour, and aroma, and consequently commands the highest value on the market.*

Cocoa is also grown in Mauritius, Madagascar, Isle de Bourbon, Australia, and the Philippian Islands.

An interesting account of the rise and growth of the West India Islands, written by Dalby Thomas, in 1690, appeared in the "Harleian Miscellany," and we extract the

---

*NOTE.—Ferguson in "Ceylon in 1884," remarked:—"Cocoa can never be cultivated in Ceylon to the same extent as Coffee, Tea, or Cinchona, for it requires a good depth of good soil and shelter from the wind, and these are only to be found in very limited areas. To the late R. B. Tytler belongs the credit of introducing this cultivation, and in his hands Ceylon Cocoa speedily realized the highest prices in the London Market, experienced Brokers remarking that there must be something in the soil and climate of the districts where it is cultivated in Ceylon peculiarly suited to Cacao. There are 10,000 acres now planted, and it is expected that ten years hence an area exceeding 30,000 acres under this plant will enable Ceylon to send 120,000 to 150,000 hundredweight of its products to European markets."

following amusing account of this early attempt and failure by the English to cultivate Cocoa:—"Cocoa is now a commodity to be regarded in our colonies, though at first it was the principal invitation to the peopling of Jamaica, for those walks the Spaniards left behind them there, when we conquered it, produced such prodigious profit with little trouble that Sir Thos. Modiford and several others set up their rests to grow wealthy therein, and fell to planting much of it, which the Spanish slaves had always foretold would never thrive, and so it happened; for though it promised fair, and throve finely for five or six years, yet still, at that age when so long hopes and cares had been wasted upon it, withered and died away by some unaccountable cause, though they imputed it to a black worm, or grub, which they found clinging to its roots." The account continues:—"And did it not almost constantly die before; would come into perfection in 15 years' growth, and last till 30, thereby becoming the most profitable tree in the world, there having been £200 sterling made in one year of an acre of it. But the old trees, being gone by age, and few new thriving, as the Spanish negroes foretold, little or none now is produced worthy the care and pains in planting and expecting it. Those slaves gave a superstitious reason for its not thriving, many religious rites being performed at its planting by the Spaniards which their slaves were not permitted to see. But it is probable that where a nation, as they, removed the art of making cochineal and curing vanilloes into their island provinces, which where the commodities of those islands in the Indians' time, and forbade the opening of any mines in them for fear some maritime nation might be invited to the conquering of them, so they might likewise in their transplanting Cocoa from the Caracas and Guatamala, conceal wilfully some secret in its planting from their slaves, lest it might teach them to set up for themselves, by being able to produce a commodity of such excellent use for the support of man's life, with which alone and water some persons have been necessitated to live ten weeks together with-

The Cocoa Tree.

out finding the least diminution of health or strength."

This inestimable plant, named by Linnæus Theobroma (from θεός and βρῶμα, the food of gods), is an evergreen which grows to the height of from 15 to 30 feet, with drooping bright green leaves, in shape oblong, eight to twenty inches long, and pointed at the ends. The flowers and fruit, which it bears at all seasons of the year, grow off the trunk and thickest parts of the boughs, with stalks only an inch long. Humboldt saw the flower bursting through the earth out of the root, and wondered at the prodigious vital force of the plant. The flowers, which grow in tufts or clusters, are very small, having five yellow petals on a rose-coloured calyx. The fruit is five-celled, without valves, from seven to nine and a-half inches in length, and three to four inches in breadth, of an elliptic oval-pointed shape, something like the vegetable marrow, only more elongated and pointed at the end, tough and quite smooth, the colour varying, according to the season, from bright yellow to red and purple. The rind of the fruit is very thick, and similar to a very hard tough apple in substance, and having a slightly sweet taste; if allowed to ripen this changes into a shell of a weak nature. The seeds contained in each pod vary in number from twenty to

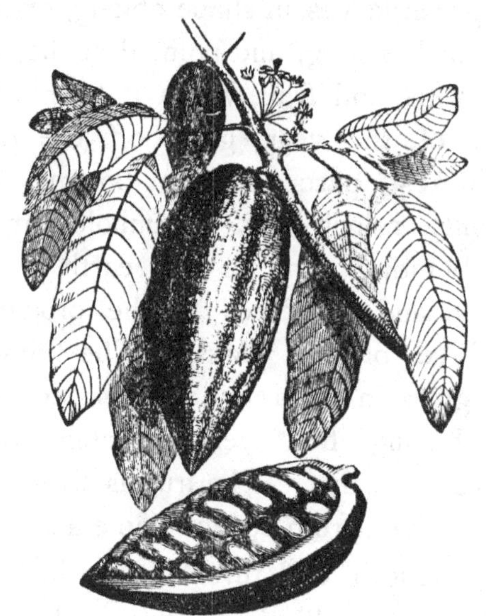

Pod, Leaves and Flower.—Pod cut open shewing Seeds.

forty, embedded in a soft pinky-white acid pulp. The cocoa tree, while growing in that portion of the earth wherein the heat is greatest, yet requires a sheltered situation for its perfection.

A writer in *Belgravia* graphically describes the beauty of a Cocoa plantation with its luscious fruit :—

"The branches do not grow low, so that in looking down a piece of ground the vista is like a miniature forest hung with thousands of golden lamps — anything more lovely cannot be imagined."

Of the varieties and cultivation of Cocoa in Trinidad and Grenada, the following extracts are from the pen of D. Morris, M.A., F.G.S., Director of Public Gardens, etc., Jamaica (1882) :—

CACAO : HOW TO GROW AND HOW TO CURE IT.

As in a large number of cultivated plants entirely propogated from seed, the Cacao plant is liable to considerable variation, even when seed is taken from the best kinds. It is important, therefore, that the Cacao planter should make himself acquainted with the chief varieties, and cultivate only those specially suited for his soil and climate—carefully weeding out, or "roguing" all those which, from habit of growth, yield, character of produce, and other circumstances are unsuitable for his purpose.

The following are some of the well marked varieties of cacao as known in Trinidad:—

1. Cacao Criollo (Red).
2. Cacao Forastero—

Var.				
a.	Cundeamor verugoso amarillo			(yellow)
b.	„ „	colorado		(red)
c.	Liso amarillo	...	...	(yellow)
d.	„	colorado	...	(red)
e.	Amelonado amarillo		...	(yellow)
f.	„	colorado	...	(red)
g.	Calabacilla amarillo...		...	(yellow)
h.	„	colorado	...	(red)

It will be noticed that the varieties are divided into two great classes, Cacao Criollo and Cacao Forastero.

Of the Forastero varieties the best are the *Verugoso Amarillo* (Yellow) and the *Verugoso Colorado* (Red). Of these two the

ALF COOKE, LITH. LEEDS

Cocoa Pods on the Branch—Trinidad.—(*Drawn from Nature.*)

yellow kind is said to yield a larger proportion of seeds than the red, and they are said to require less time for fermentation. It appears that in Grenada also the yellow is preferred to the red, both on account of its yield and its greater adaptability to the exigencies of cultivation.

The average mean temperature of the plains and valleys of Trinidad are naturally higher than similar localities in Jamaica, but if we would compare the temperature of Jamaica with that of the plains and valleys on the coasts of Guatemala and Mexico it would be found that our temperature is quite high enough for the successful cultivation of the Cacao plant.

As regards elevation, with the exception of some mountain ranges in the North, rising into sharp peaks of about 3,000 feet, the surface of Trinidad is in general flat, or gently undulating. The highest cultivated lands seldom exceed 200 feet or 300 feet, and no parts are inhabited above 500 feet. In Arima and the Monserrat districts, the Cacao estates occupy open glades and moderately undulating country, ranging from almost sea level to 150 feet or 200 feet. The sub-soil in these districts is of a marly character, overlaid, chiefly in hollows, by gravelly loam, moderately deep and remarkably free from rocks and stones.

In Grenada, Cacao is generally cultivated at a higher elevation than in Trinidad, some estates occupying hill slopes up to 800 feet. The best estates are, however, at the foot of the hills or in sheltered glades, on land formerly cultivated in sugar.

As regards the planting, Cacao requires more care and thought than is generally imagined.

For instance, it is not only necessary to be ready before-hand with Cacao seeds or plants, but the plants have to be protected by larger trees necessary for shade, and put in either before the Cacao or at exactly the same time. The Cacao are planted at exactly the same distances apart, occupying the centres of

Curing House in Grenada.

squares. Close to the Cacao plants are small

Copy of an old engraving (said to be the oldest in existence on Cocoa) showing the way in which the Cocoa Tree is shaded by other larger trees.
From Boutekoe's Works.

shade plants to protect them for a few months: further off are the bananas and plantains, one between each Cacao plant, to last for about two or three years; and lastly there come the permanent shade trees, at distances of 39 or 40 feet, which at the end of three or four years will be the only occupants of the ground besides the Cacao. It delights in a deep and moderately rich soil—preference being given to that containing a certain proportion of lime or marl.

In planting, either of two systems may be adopted :—

1—(*a*) Planting at stake (seeds), or (*b*) planting from nurseries (plants).

Planting at stake :—In this case the best and largest seeds in a pod are taken and two or three are planted at each stake, the soil being first softened and broken up by a hoe.

After being sown, the seeds require both shade and protection till they have germinated, which they generally do in a week or ten days. When the young plants are from four to six months old, the strongest only is retained, the others being carefully removed to give it full scope to grow.

Planting by stake is only adopted in fresh good land and where seeds are abundant.

In addition to selecting the best and largest seeds (leaving out the end ones) it is advis-

Alf Cooke Lith Leeds

Cocoa Plantation,—Trinidad.—(*Drawn from Nature.*)

able for this purpose to wash the pulp and cover them thoroughly with wood-ashes, as a protection against ants and predatory vermin.

Planting from Nurseries:—Where planting at stake is not practicable, it is advisable to establish, beforehand, nurseries raised from seed of the best varieties, so as to have plants ready for putting out with the first rains. If the number be small it would be better to raise the plants in bambu pots, as well for convenience of transport as for protection to the young plants in the process of transplanting.

Cacao trees in good situations begin to bear in about the third or fourth year. Individual trees will, however, sometimes show fruit when only two years old, but it is much better for the trees themselves that they should be stripped and not allowed to bear till at least the fourth or fifth year. A Cacao plantation should be in fair bearing from the sixth to the ninth years, and at its prime from the twelfth year.

Gathering Crop:—Although Cacao is in bearing more or less all the year round, the chief crop seasons are in May and June, and again in October and November—these are known in Venezuela, where the famous Caracas Cacao is grown, as the St. John's and Christmas crop, respectively.

Gathering crop is done as follows:—A number of men, each supplied with a long bambu rod surmounted by a Cacao hook and a cutlass, go carefully over the plantation and pick out all the ripe pods.

These are known by their colour, or better still, by tapping them. If ripe they give a hollow sound, as the seeds are then loose and detached from the outer shell. In gathering the higher pods the Cacao hook is used, but the lower ones are taken off by a cutlass. The Cacao hook is constructed so that it will sever the pod either by a thrust or by a draw.

Few operations upon a Cacao estate require greater care than gathering crop, and for the following reasons:—

At the place where a pod is attached (formerly a leaf axil) there is a soft cushion or "eye," from which all subsequent flowers and fruits arise. If this "eye" be damaged—as it inevitably would be if the pod were ruthlessly torn off instead of being cut—the tree, as far as this point is concerned, becomes sterile. Hence, if a succession of these "eyes" are thus treated the tree would ultimately become practically valueless.

When the pods have been gathered and left in small heaps near the trees, they are collected by women into larger heaps, and left till the next day.

The larger heaps are generally placed near a clear, open space, where the processes of "breaking" and "drawing" can be conveniently carried out. It is advisable, however, not to use the same spot too often, as otherwise the empty pods accumulate and prove an impediment to the cultivation.

A party, consisting of a man with a cutlass

and two or three women with wooden spoons

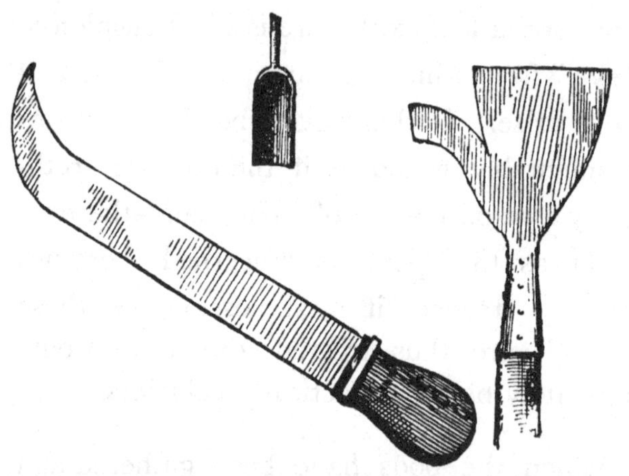

Implements used for severing the Fruit from the Tree, and cutting open the Pod.

or scalpels, are told off to a number of heaps, and by each one they spread plantain leaves on the ground to receive the seeds. Then, while the man breaks the pods with his cutlass, the women remove the beans with the wooden spoons, clean them of the fibrous tissue by which they are attached, and throw them in a large heap.

All black, unripe, or damaged beans are placed on one side.

When this process is completed, the fresh

seeds are ready to be conveyed or "crooked" to the Cacao house, and placed in the "sweating" house.

Sweating House, Grenada

This process is one upon which, in a great measure, the commercial value of Cocoa depends. The first impulse of a grower would be to remove the pulp between the seeds by washing, and dry the beans as quickly as possible. This, however, would not result in a good quality of Cocoa. The pulp must be removed, but washing is not the best process.

The fresh bean will be found to be somewhat bitter in taste, and a pale crimson colour;

both these have to be altered before the bean is fit for the market.

The best means for this purpose is evidently the one now generally adopted in all good Cacao growing countries, and that is "sweating" or fermenting the beans. This, for the most part, gets rid of the pulp, softens the bitterness of the fresh beans, and gives them, when cured, that rich mahogany tint so much sought for by chocolate makers.

The sweating process may briefly be described as follows:—The beans brought from the field are placed either in barrels, oblong boxes, or in a close room, where they are packed closely together covered with plantain leaves, and left hermetically closed for a period extending from four to seven days. The exact number of days will depend on the variety of the bean or quality of Cacao desired. While thus shut up, a process of fermentation, fed by the saccharine matter in the pulp, takes place, which raises the temperature of the mass to about 140° Fah. During fermentation carbonic acid is given

Collecting the Cocoa.—(*Drawn from Nature.*)

off, and some water. In wet weather care is taken that the temperature of the mass does not rise too high, as otherwise the beans would blacken. It is often necessary, under these circumstances, to open the Cacao, and carefully stir it before it is returned, to complete the fermenting process.

For a plantation, say above 10 acres, it would be more convenient and satisfactory to have a small building for the purpose attached to the Cacao house, called a "sweating house."

This house would be somewhat as follows: An oblong room on the basement story, or mounted on pillars, with boarded sides carefully fitted, so as to be perfectly air-tight. The only opening into it is by a door, which should also fit as tightly as possible. For keeping crops gathered at different times distinct, it is advisable to have a division in the middle. The floor should be double, and constructed of rather close parallel bars, so as to allow water, but no beans, to fall through into a space below.

When the formerly pale crimson colour has given place to a brownish tint, the Cocoa is turned out and spread on the "tray" or "barbecue." It is first of all carefully picked over by women, who separate the beans from "trash" or any foreign substances. This done, the beans are covered with red earth, and left to complete the process of fermentation for another day. A number of women are then employed for one or two hours in rubbing them with their hands, and cleaning them as thoroughly as possible from all mucilaginous and gummy matters.

The red earth, by its absorbent qualities, assists in ridding the beans of the mucilage, and gives them a deep red colour; it is also supposed to give them better keeping qualities. A large proportion of Trinidad Cacao is cleaned without the use of red earth, but the process is much more tedious and the beans are not so good in colour and general appearance. In Jamaica, a large quantity of red earth for the purpose can be obtained from what are called the "Red Hills," St. Andrews.

When the cleaning and rubbing process has been completed, the beans are spread out on a tray to dry.

While drying they are carefully turned, so as to expose all the beans to the influence of the sun; but in case of rain they are immediately covered by the sliding roof of the

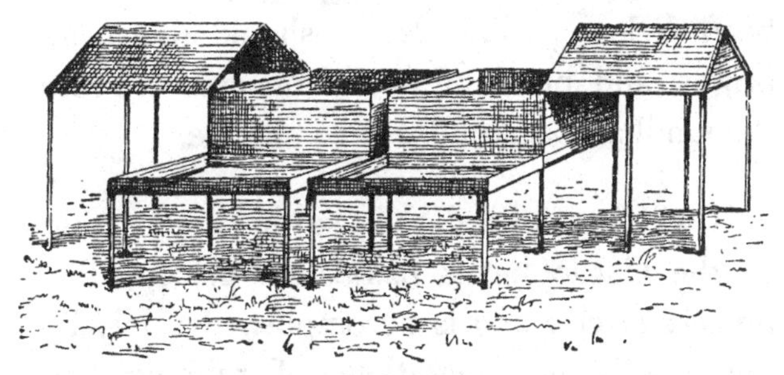

Trays for Drying the Cocoa.

Cacao house. During the hottest part of the day, when the thermometer stands over 90° in the shade, it is considered undesirable to expose the beans too much to the sun, as thereby they become "parched" or shrivelled. The Cacao house is therefore generally shut on hot days for three or four hours, and the Cacao exposed only when the temperature is low.

The process of turning and drying is continued from day to day until the Cacao is thoroughly cured.

The experience of the planter alone can tell whether the Cacao is thoroughly dry.

If well cured it should have the outer skin hard, crisp, and separating easily from the bean below. The latter should be firm, bright, and breaking easily on pressure into the familiar Cacao nibs of commerce.

It will be noticed that, so far, no washing of the Cacao beans has taken place. The process of cleaning is accomplished solely by the alternating operations of rubbing and drying with, as already mentioned in some instances, the aid of red "clay" or earth.

A good Cacao tree in good soil yields from fifty to several hundred pods per annum. The average for well-cultivated trees, at seven years old, should be between eighty and one hundred pods per annum. As it generally takes about eleven pods to yield one pound of cured Cacao, the above would indicate that a good mature Cacao tree, under favourable circumstances, might yield, on an average, not less than seven pounds of cured Cacao. The average yield per tree (at all stages) on an estate of, say 300 acres, would probably not exceed some two or three pounds per tree, or (taking 230 trees per acre) a return of 4 cwt. to 6 cwt. of cured Cacao per acre.

CHAPTER II.

——:o:——

HISTORY OF THE USE OF COCOA.

——:o:——

OOKING at the early works on the use of Cocoa we find them very quaint and speculative as to its medicinal value, and its effect upon the constitution.

As early as 1624 Joan Franz Rouch wrote a treatise condemnatory of its use, and at the same time abusing the monks.

About contemporary with this was a book written by Antoino Colmenero de Ledesma, Medecin and Chirurgien, de la Ville de Ecija, de l'Andalouzie," 1631; this was translated from the Spanish into French by Renè Moreav, in 1671; and into Latin by Marco Aurelio Severino, in 1644.

Copy of an engraving, from a rare work by Philippe Sylvestre Dufour, showing a native with his chocolate pot and drinking cup, and the "molinet," or stirrer, in his hand.

Willem Boutekoe, a Dutch author and traveller, wrote sundry short treatises on Cocoa and Chocolate about 1679. De Chélus, 1719, wrote an "Histoire Naturelle du Cacoa et de Sucre."

Another French work "on the quality and nature of Chocolate," by Philippe Sylvestre Dufour, in 1688, from which we copy some of the very interesting engravings. The first work we have seen in English being "Translated from the last edition of the French by R. Brookes, M.D.," 1730.

Our knowledge of Cocoa as an article of diet dates from the discovery of the Western world by Columbus, in 1494, who, we are told, took home with him samples of the article; and the subjugation of Mexico by Cortez in 1521. History informs us that the Spaniards were the first who tasted Chocolate, which was part of their spoil in the conquest of Mexico. Bernardo de Castile, who accompanied Cortez, describing one of Montezuma's banquets, says:—"They brought in among the dishes above fifty great jars made of

'Cacao,' with its froth and drank it," similar jars being served to the guards and attendants "to the number of 2,000 at least."

Prescott, in his history of the Conquest of Peru, says that "The emperor took no other beverage than the chocolatl, a potation of Chocolate flavoured with vanilla and other spices, and so prepared as to be reduced to a froth of the consistency of honey, which

Copy of an engraving from Dufour's book.

gradually dissolved in the mouth, and was taken cold. This beverage, if so it could be called, was served in golden goblets, with spoons of the same metal or of tortoiseshell finely wrought." The historian also adds, "the emperor was so fond of it, to judge from the quantity, no fewer than 50 jars or pitchers being prepared for his own daily consumption, whilst 2,000 more were allowed for that of his household."

Mendoza, in his work upon the "Antiquities of Mexico," tells us that amongst other things paid into the Mexican Treasury as tribute by different cities of the Empire, there were "20 chests of ground Chocolate, 80 loads of red Chocolate, and an item of 200 loads of Chocolate. We also find that the Cocoa seed were used by the Mexicans for currency. Peter Martyr gave them on this account the name *Amygdalæ pecuniariæ*.

It may be as well to explain here that the word "Chocolate" is of Mexican origin, being derived from "chocolatl"; the pronunciation of the word resembling the clattering sound produced by the native handmill used to grind the Cocoa and mix it with sugar.

Thomas Gage, in his "New Survey of the West Indies" (1648), says "the name is compounded from *atte*, as some say, or as others, *atle*, which in the Mexican language signifieth water, and from the sound which the water (wherein is put the Chocolate) makes, as *choco*, *choco*, *choco*, when it is stirred in a cup by an instrument called a 'molinet,' or 'molinillo,' until it bubble and rise unto a froath."

Chocolate Stirrer (Molinet), copied from an old book published in the 17th century.

The same writer remarks: "Our English and Hollanders make little use of it when they take a prize at sea, as not knowing the secret virtue and quality of it for the good of the stomach." For many years the cultivation of the Cocoa tree was confined to the Spanish, who, in South America and some of the West India Islands, carried on the cultivation to a large extent. From their first settlement in Trinidad, we are told that "it seems probable the Spaniards cultivated the tree, and thoroughly understood its value, the prepared article being always much esteemed in the then opulent mother-country."

From Spain the monks introduced the use of Chocolate into France, in 1661. The earliest record we have of the use of Chocolate in England is said to be furnished by an advertisement which appeared in the *Public Advertiser*, or *Adviser* according to one authority, of Tuesday, June 16, to Tuesday, June 22, 1657, informing the public that "in Bishopsgate Street, in Queen's Head Alley, at a Frenchman's house, is an excellent West India drink called Chocolate to be sold, where you may have it ready at any time, and also unmade at reasonable rates."

Disraeli, in his "Curiosities of Literature," speaking of the introduction of tea, coffee and chocolate into Europe, says: "Chocolate the Spaniards brought from Mexico, where it was denominated *chocolatl.* It was a coarse mixture of ground Cacao and Indian corn with *rocou;* but the Spaniards, liking its nourishment, improved it into a richer compound with sugar, vanilla and other aromatics. We had Chocolate-houses in London long after coffee-houses; they seemed to have associated something more elegant and refined in their new term when the other had become common."

Cocoa was much esteemed as a beverage in this country during the reign of Charles II., and at that period Dr. Stubbe published a book entitled "The Indian Nectar, or a Discourse concerning Chocolate, &c.," in which the author gives a history of that article, and many curious notions respecting its "secret virtue," and recommends his readers to buy it of one Mortimer, "an honest though poor man," who lived in East Smithfield, and sold the best kind at 6s. 8d. per lb., and commoner sorts at about half the price. It was not until the close of the sixteenth century that Cocoa or Chocolate was generally used in this country, and when we take into account the indifferent means for preparation and the adulterated condition of the article we can hardly be surprised that it did not come into general favour with the public.

Reproduction of an old Plate (1688) from Dufour's Treatises on Coffee, Tea, and Chocolate

Prior to 1831, the quantity consumed in this country only amounted to 500,000 lbs. per annum.

Sixty years has wrought a marvellous change, which may best be described by the following facts and figures of comparatively recent date. In 1858 only 2,860,034 lbs. of Cocoa were consumed in this country; in 1864, 3,862,273 lbs.; in 1880, 10,566,159 lbs.; and in 1890, 20,224,175 lbs.

Of the Cocoa now imported into England, close on one-third is cleared by the house of Cadbury, Bournville, an account of whose factory we give in another chapter.

The consumption of Cocoa was quite nominal in the earlier part of the century, 1s. 6d. per pound being levied for duty up to 1820, and every pound of Chocolate had to be wrapped in papers supplied by the revenue officer.

In an article by Dr. A. J. H. Crespi, Wimborne, in *The Housewife*, December, 1889, he says:—

The consumption of tea is now said to stand at 150,000,000 lbs. a year or more, and of coffee at 80,000,000 or 100,000,000 lbs., while Cocoa still only reaches 18,464,164 lbs., or, roughly speaking, eight ounces per head, a surprisingly small consumption, especially when it is remembered that so much of it goes in bon-bons; indeed, many people never drink a cup of Cocoa or Chocolate from year's end to year's end, though they get through, whenever they have the opportunity, a large quantity of Cocoa or Chocolate sweetmeats. We see no reason why the average consumption of Cocoa should not stand at one pound a head, that is at nearly 40,000,000 lbs., for of late the growth of the trade shows marvellous improvement, and our conservative countrymen—conservative for good as well as for evil—have at last begun to recognise the value of Cocoa as a household beverage of a class with absolutely no other member.

The chemical composition of Cocoa as given in a paper read before the Society of Arts by John Holm, F.R.G.S., in 1874, is as follows, and the table contains all analyses of any importance made up to that time:—

ANALYSES OF COCOA.

	Lampedius.	Payen.	Johnson.	Playfair and Lankester.	Miller.	Mitscherlich.		Muter.	Average of several other analyses.
Fat (Cocoa Butter)... ...	53·10	52·00	51·00	50·00	56·00	45·00	49·00	42·67	50·00
Albuminoid Substances ...	18·70		...	...	17·00	13·00	18·00	...	...
Albumin	...	20·00	...	20·00	...	...	...	...	18·00
Fibrin	...		...	...	...	...	...	...	
Gluten	...	...	20·00	...	...	...	...	12·21	...
Extractive Matter	...	...	...	...	...	...	...	...	...
Sugar	...	...	...	...	...	0·60	...	...	...
Starch	10·91	10·00	22·00	7·00		14·00	18·00	19·03	10·00
Gum	7·75	...		6·00		...	...	6.40	8·00
Lignine	0·90	...	...	...	22·00	...	...	...	...
Cellulose	...	2·00	...	...		6·08	...	5·95	...
Woody Fibre	...	...	...	4·00		...	...	...	...
Colouring Matter	2·01	traces	...	2·00	...	3·05	5·00	3·96	2·60
Water	5·20	10·00	5·00	5·00	...	5·06	6·30	5.98	6·00
Theobromine	...	2·00	2·00	2·00	1 50	1·02	1·50	0·90	1.50
Salts	...	4·00	...	4·00	...	...	...	...	...
Ash	...	...	...	...	...	3·05	...	2·90	3.60
Humic Acid	...	...	...	...	...	...	...	...	...
Parts unaccounted for ...	1·43	...	...	...	3·50	9·14	...	...	0.30
Total	100·00	100.00	100·00	100·00	100·00	100·00	...	100·00	100·00

Mr. Holm also makes the following observations:—

It is a table which is not very flattering to chemical science, the analyses being of the most contradictory character, and containing discrepancies which cannot be at all reconciled with each other. I should judge that the analysis prepared by Drs. Playfair and Lankester is the most correct. We thus see that, taking the important constituents, Cocoa contains:—

	Parts.
Cocoa butter	50
Albuminoid substances	20
Starch, Sugar, &c.	13
Salts	4
Theobromine	2
Other constituents	11
	100

Taking these in the order of their importance, we first notice the fat, or Cocoa-butter, forms about half the substance of the nibs. It is a hard, fatty substance which, when clarified, is of a pale yellow colour. Its melting point is about 100° Fah., which being the heat of the body, renders it of great value for therapeutical purposes. The fat only becomes rancid when subjected to heat or light, and especially to the direct rays of the sun. It is hardly necessary to point out how valuable this quality renders this portion of the bean for surgical and other purposes. The albuminoid constituents form about 20 per cent. of the nib. These are classed amongst the nitrogenous principles of food and their presence renders Cocoa one of the richest flesh-formers we have. The starch, gum and sugar present, like the Cocoa-butter, belong to non-azotised principles ; they form about 13 per cent. of the whole. The alkaloid of Cocoa, *theobromine*, is very similar in its physiological effects to its analogues, *theine* and *caffeine*, from which it differs very slightly in chemical composition.

ESSENTIAL ALKALOID PRINCIPLES.

Yielded by	Name.	Composition.	Proportion.
Cocoa	Theobromine	$C_7 H_8 N_4 O_2$	2 per cent.
Coffee	Caffeine	All Identical $C_8 H_{10} N_4 O_2$	1 to 7 per cent.
Tea	Théine		
Guarana	Guaranine		
Maté	...		In less quantities

In regard to these alkaloids it is interesting to note that throughout the world the instinct of man has led him to seek some substance which contains one of these principles, which owe their value to the specific influence they exert on the nervous system, stimulating it and checking waste of tissue. *Theobromine*, when extracted, presents the form of a white crystalline powder of almost amorphous appearance, differing from *caffeine* and *theine*, which have a very beautiful crystalline appearance.

In most of the analyses of Cocoa the existence of a volatile oil has been overlooked. It is probably present only in small quantities, and appears to be developed by roasting; but upon it depends the flavour and aroma which exists in Cocoa.

CHAPTER III.

—:o:—

MANUFACTURE.

—:o:—

E cannot do better than transport our readers to Bournville, " the Worcestershire Eden " as it has been aptly termed, for the surroundings of the place have a charm of their own that banishes the thought of smoke and machinery, and gives quite a zest for the inspection of this happy and busy scene of labour. Bournville is certainly a model factory, both for its size and its completeness, and because it contains the most modern improvements in the application of machinery for the manufacture of Cocoa and Chocolate.

Five tall chimneys and some acres of warehouses and workshops divided by streets connected by bridges, and intersected by railway lines, give some idea as to the extent of the works, but for all this our first impression on being taken through the porter's lodge was that of entering into a garden with a welcome of the sweet breath of flowers and the song of birds; for we have to pass by the girls' garden and play-ground, which is well planted and bordered on three sides by trees and shrubs, while the plantations of Bournbrook Hall give the substantially country surroundings of which it boasts. One of the sights of Bournville is to see the girls in their white costumes, after dinner, sitting to read under the trees, or enjoying in other ways a breath of fresh air.

The name "Bournville" was suggested by the immediate contiguity of the pretty streamlet known as the "Bourn," which meanders through the estate and forms the northern boundary of the factory. This rippling rivulet adds to the attractiveness of the

Girls' Playground and Waterfall at Bournville.—*(From Photographs.)*

locality, and is regarded with no little satisfaction by the disciples of Izaak Walton, who recognise in the pretty trout stream an important feature which often gladdens the heart of the angler. Among those engaged at the works are lovers of the enthralling pastime, and we are informed that recently trout of very respectable dimensions had been landed from the Bourn.

On ground contiguous to the factory, and bordering on the road which runs on the southern side of the works, Messrs. Cadbury have built 16 semi-detached villa residences, which are inhabited by their most prominent hands—workmen who have shown by their general demeanour, diligence in business, and assiduity, that they are worthy the consideration their employers have evinced for their interests. These villas are model residences, let at a comparatively small rental; they are well built of brick, in two colours, fitted internally with taste, and each would easily bring a rent of £40 per annum in the suburbs of the metropolis. The inhabitants of these residences pay 5s. to 6s. per week. To each house is apportioned a front and back garden, and the tenants are enabled to grow their own vegetables and fruit, and decorate the front parterres with flowers.

Like many of our largest manufacturers, Messrs. Cadbury commenced business with a staff of workpeople comparatively small when compared to the number of hands at present employed by them. About thirty years ago under twenty employés comprised the working establishment, while at present they employ about 1,600 men, boys, and girls.

Owing to the comparatively isolated position of the works, ample provision has to be made for all requirements as regards cooking. Spacious dining rooms have been provided separately on the premises for both men and women. Gas stoves and cooking apparatus have been erected, and hot dinners can be procured in a very few minutes. So complete are the cuisine arrangements that there is little delay in serving all from the

Workpeoples' Cottages and Country Lane at Bournville.—(*From a Photograph.*)

kitchen, which is constructed between the men's and women's dining rooms, which are kept quite distinct.

The manufacture of Cocoa and Chocolate requires great experience, skill, and special knowledge. In detailing the processes it will be convenient to divide them into two branches, viz., the production of Cocoa Essence, and the manufacture of sweet Chocolate.

Cocoa Essence, which is the speciality of this firm, is unrivalled as a nutritive beverage, and therefore the most important for consideration. The best Cocoa contains about 50 per cent. of natural oil or butter, and this has been found to be far too large a proportion for ordinary digestions. Dr Muter remarks that the "only objection which can and does exist to its use in a state of purity is the excessive proportion of fat, which renders it too rich for most digestions, and gives, unfortunately, a colourable excuse for its adulteration.'

By means of elaborate machinery at these works the removal of two-thirds of the butter is accomplished, the result being an impalpable powder easily miscible in boiling water.

As the visitor leaves the office of the principals, and enters the factory, the fragrance of the Cocoa-berry salutes the olfactories most agreeably. The hum of machinery denotes that the hands are busily engaged, and as we pass into the factory, an earnestness of purpose is manifested by the workpeople who are intent upon their various pursuits. The utmost order and regularity is preserved in all the departments, and every employé appears to discharge his or her duty with that ease and readiness which is the result of experience and training.

The Cocoa arrives in sacks weighing from one to two cwt., and as it varies considerably in kind and quality, it has to be stacked in large piles ready to undergo the first process of sifting and picking, so that no unsound berry or other foreign material is passed into the roasting room. The sieves used for this

ONE OF THE ONE-TON STEAM ROASTERS AT CADBURY'S WORKS

process are long barrels on a slight incline, which slowly revolve and sort the nuts into various sizes, while at the same time they remove dust or smaller matter that may come with them. By an automatic process the nuts are carried into the hoppers of the roaster, three of which are each capable of roasting one ton at a time. These rotate slowly and are roasted by high pressure steam, being especially adapted for a particular purpose of manufacture.

There are two other ways used of roasting Cocoa, by more direct and intense heat, and which is of course a quicker process than by high pressure steam. So important is this process that very careful attention is necessary, and experienced workmen, whose judgment is almost unerring, are entrusted to superintend the roasting. It is requisite for those who have charge of this department to determine the precise period at which the nuts are sufficiently roasted, for the quality and rich aromatic flavour of the Cocoa depends greatly upon this. A miscalculation in time would tend to spoil the Cocoa, but it is satisfactory to know that mishaps rarely happen, and so practised are the hands responsible for the roasting that the work is, as a rule, admirably done and the flavour of the nut is invariably preserved. In connection with this process, methods of treatment peculiar to the establishment are successfully adopted. After being roasted the nuts are placed in trays of considerable superficial dimensions to cool. The fresh air speedily reduces the temperature of the Cocoa-beans, and they are then ready to be what is technically termed "broken down." The now crisp roasted nuts are placed in a hopper and afterwards raised by an elevator, and passed through a machine which gently cracks them, disengaging the hard thin skin, which by this means can be separated from the nutritive portion of the nut, viz., the rich glossy kernel, known in the market as Cocoa-nibs. The separation is effected by a winnowing machine. From the outlet of the cracking machine the husk and nut are carried to a point over the winnower, and as the cracked nuts fall, the powerful blast

GRINDING MILLS
AT CADBURY'S WORKS
(FROM A PHOTOGRAPH)

of this machine blows away the husk from the nut, and the latter falls into a receptacle in the form of nibs, which are sorted by a diviseur. The husk or shell is sent off to Ireland and elsewhere to be used as a light, but by no means unpalatable, table decoction, under the designation of "miserables."

The mill room, into which we now pass, is a very spacious and well-arranged apartment, in which numerous machines are employed in the manufacture of Cocoa and Chocolate, the most approved modern mechanical appliances having been introduced. In this room three long lines of millstones are at work crushing the nibs, which are fed into a hopper, from whence they pass between granite millstones. As these stones are heated the nibs are reduced to a creamy fluid, which flows into a receptacle. The nibs are hard and brittle before they are crushed, but after a few minutes grinding the oil they contain is disengaged by the heat, and an oleaginous paste is produced. From this fluid the Cocoa-butter is extracted by means of a certain process—a speciality of the firm—and the substance is left perfectly dry. This is speedily reduced to an impalpable powder, and the well-known and absolutely pure Cocoa essence, for which the firm is so celebrated, is complete.

It should be explained here that there are three forms in which absolutely pure Cocoa can be used, namely: 1st, the Cocoa-nib before grinding into a paste, prepared for drinking by making an infusion from them by boiling in water; 2nd, the Cocoa-nib ground into a paste and solidifying into a hard cake and retaining all the butter: this is not soluble in boiling water; 3rd, the Cocoa essence, which is practically soluble, containing a larger proportion of flesh-forming substance.

Many millions of mill-board boxes, to contain the Cocoa essence, have to be made yearly, and it may therefore be readily supposed that the box-making department is one of considerable importance. The demand being so enormous the firm employ elaborately-constructed machinery to meet their requirements. One machine cuts the

A Quiet Corner at Bournville.—(*From a Photograph.*)

board into the required shape, while another glues the parts together and perfects the packet - shell, the output averaging about 5,000 daily from each of the six machines. They work with perfect regularity, and demonstrate the state of perfection to which labour-saving inventions have been brought. The boxes are removed by means of a hollow band, and forwarded to the packing-room, where numbers of busy hands are filling, wrapping, and labelling the packets.

We now come to the second branch of manufacture, which is of considerable commercial importance, and very extensive. Sweet Chocolate, for eating and drinking, forms the most delicious of all confections or beverages, and Cocoa prepared in this way is another of the specialities of the firm. To the manufacture of the numerous varieties this last section of the factory is devoted. The pure Cocoa is, in the first place, incorporated with white sugar in what is called a

"Melangeur." This mixing machine consists of a round granite revolving slab, forming a pan, the sides being of steel. Into this receptacle the Cocoa and sugar are poured, and two sets of heavy stationary granite rollers bruise the thick mass, which is reduced to the consistency of dough. A double knife, the action of which is similar to that of a screw propeller, continually revolves just above the rotary stone slab, and distributes the chocolate as it passes. There were several of these machines at work, and our attention was particularly directed to one of an improved design and great magnitude, which mixed and ground the Chocolate, and then automatically passed it on to heavy granite cylinders, which systematically and gradually reduce the chocolate to a given degree of fineness, the operation being effected with remarkable exactitude.

One special article made by the firm has been compared to the famous Chocolate that Prescott describes as forming part of Montezuma's repast—"In golden goblets flavoured with vanilla, and so prepared as

LARGE MÉLANGEUR, WITH EIGHT HEAVY GRANITE CRUSHING ROLLS, AT CADBURY'S WORKS.

to be reduced to a froth of the consistency of honey, which gradually dissolved in the mouth."

In this establishment the charm said to be inseparable from variety is not lacking. The creme moulding-room presented a scene that was unlike any previously witnessed within the works. The work upon which the young women in this room were engaged was of a delicate and light description, particularly suitable for female hands. In this portion of the factory the delicious Chocolate creams, which we need not describe because they are so well known, are poured rapidly into moulds of various patterns and designs. These moulds are formed in finely prepared corn flour, which gives the room the appearance of a flour store ; the workers dexterously pour the liquid cream along the mould, and each hole absorbs the alloted quantity. The young women engaged at this work are remarkably expert and skilful, as they manage to fill each mould with just sufficient of the cream and no more, thereby preventing waste of time, if not of material. The creams soon solidify, and when cool they are extracted from the moulds, and the flour in which they were moulded having been brushed away, they are taken to another department and coated with Chocolate.

The room which we next enter is very extensive and commodious, being 240 feet in length by 60 feet wide. Every business convenience is afforded in this noble workroom, and the young women, with ample space at their command, suffer under none of those disabilities which invariably arise when the area is insufficient and the apartment is "cribb'd, cabin'd, and confin'd." A large number of young women are engaged in this and other rooms, of slightly less proportions, boxing, labelling, and making fancy boxes to contain Chocolate creams and numerous confections composed of the delicious preparations so well known and highly appreciated.

Not only is it essential that the sense of taste should be gratified by the manufacture of toothsome compounds but the eye must be pleased. Many varieties of fancy boxes made in these rooms are admirable examples of art workmanship. The designs and pictures on some of the best packages are chaste and elegant, while in the vast assortment of decorative embellishments every taste and fancy may be gratified. At Christmastide boxes and creams are in universal request. Passing around this department an interesting and diversified scene presents itself. Young women are busy at work at their tables or counters, some being engaged in cutting out and stamping the cardboard, and others fitting the boxes together on blocks, wrapping the edges with gold or gelatine paper, and fixing on the top those pictorial artistic gems which are sc attractive. Many others are employed ornamenting and finishing, with exceedingly good taste and evident skill, the different sorts of boxes required for the various goods manufactured at the works. The hands have ample material at their disposal to render their work effective, and every novelty likely to please the public is promptly introduced. Some of the latest and most artistic ornamentations well deserve a frame.

In other rooms numerous mechanical appliances are used for cutting cardboard, paper, &c., and the number of fancy boxes turned out weekly is enormous. In another part of this section young women were busy packing creams in boxes, and wrapping Chocolate cakes in tinfoil and papers of various colours. Many thousand of completely-finished boxes and packets of every description, containing Chocolate creams, plain chocolates, and in fact, almost every variety of the firm's manufacture, including the specialities already mentioned, were ready for packing and transit to all parts of the world, to supply the home, foreign, and colonial markets.

The saw-mills and wood box-making department are distinct portions of the establishment, and the visitor on entering these extensive workshops would imagine that another industry, entirely removed from the manufactory of Cocoa and Chocolate, was being pursued. In this building there is a buzz and whirl, caused by the circular-saws, by which spruce-planks are reduced to the required length and thickness for box - making, and most efficient planing-machines renders the surface of the wood as smooth as glass. These are transferred by a lift across the road to larger premises, where a number of hands are employed to nail the pieces together by ingeniously contrived machines which punch in three nails by one process, and through the remarkable activity of the workers, a box is nailed together almost before you can see how it is done.

The tinman's shop is next in rotation. Ingenious tools and appliances of modern design and construction have been introduced, and the tinwork, which is considerable in so vast an establishment, is executed with skill and dispatch by first-class workmen. Many thousands of tin-boxes for packing the Cocoa essence are turned out every day, and moulds for the chocolates are made and stamped, the zinc linings for export cases, &c.; also general repairs incidental to the works, which come within the province of the tin-worker, are done in this department, which is fitted with every requisite contrivance for economising labour and securing satisfactory results.

We may state here that the most excellent system of payment is adopted, by results. At this factory almost all the employés are engaged in what is known as piece-work, which is satisfactory alike to the firm and to the hands. The system has, we understand, worked well, and the work-people, both male and female, are well satisfied with the manner in which they are treated, and we were gratified to be assured more than once how fortunate it was deemed to be employed at Bournville. This good feeling between employers and employed is of almost inestimable value, both socially and commercially.

It would be foreign to our purpose to describe fully all the interesting details of such an establishment, but we may remark in closing that if all manufacturers would make the interests and happiness of their employés a part of their business, it would add to their prosperity and do something to solve the important problem of labour by cementing the friendship of masters and workpeople.

CHAPTER IV.

—:o:—

VALUE OF COCOA AS FOOD, AND ITS ADULTERATIONS.

—:o:—

"'Tis not enough to help the feeble up,
But to support him after."

TIMON OF ATHENS.—*Act I., Scene I.*

N a passage from one of Froude's charming "Short Studies" he says: "Observe the practical issue of religious corruption. Show me a people whose trade is dishonest, and I will show you a people whose religion is a sham." "We have men that steal money," Erasmus exclaimed, writing doubtless with the remembrance of a stomach-ache. "These wretches steal our money and our lives too, and get off scot-free."

Keen observers of the national progress cannot have failed to notice the growing interest taken in all questions relating to the three essentials conducive to health and longevity, viz., wholesome food, pure water, and fresh air. In these progressive times there is a spirit of inquiry and investigation manifested, and the consumer is no longer content to take things as they are, but, on the contrary, being of an "inquiring mind" he is desirous to ascertain, for his own satisfaction and benefit, "what to eat, drink, and avoid." It is well known that adulteration was, in past days, carried on to a very considerable extent, and although very much has been done to mitigate the evil, "the selling of an inferior or debased substance under the name of a superior or genuine article" still continues, both as regards food and drink.

All foods are classified into two types or divisions—namely, the nitrogenous, or tissue-forming, and the calorificient, or heat-creating.

By the term "nitrogenous" is meant all foods, whether derived from the animal or vegetable kingdom, which contain nitrogen as one of the elements of their composition, in addition to carbon, hydrogen, and oxygen. These foods are also called tissue-formers, and the measure of their flesh-forming value is the quantity of nitrogen they contain. The reason of this is found in the fact that all the tissues of the body, fat excepted, contain nitrogen, and those wherein the nutritive changes are most active, such as muscle and nerve, contain the largest amount of nitrogen.

The active principle of Cocoa is Theobromine, of which active principle we find, according to Drs. Playfair and Lankester—

Tea contains	... ...	3	per cent.	Theine.
Coffee „	... ...	1¾	„	Caffein.
Cocoa „	... ...	2	„	Theobromine.

Cocoa also contains a volatile oil, which gives its delicious aroma, and, no doubt, essentially adds to its refreshing and exhilarating character as a beverage.

Nearly nine-tenths of the Cocoa bean is composed of matter that is assimilated by the digestive organs ; while with Tea and Coffee more than one-half is thrown away as waste product. The proportions of woody fibre are as follows :—

Tea	20 per cent.	Woody Fibre.
Coffee	35 „	„
Cocoa	4 „	„

Cocoa is said to yield thirteen times the nutriment of Tea for the same value, and four-and-a-half times as much as Coffee.

The importance of these facts in connection with the use of Cocoa will at once be apparent when we compare the analysis of Cocoa nibs, which contain all the natural butter, with Cocoa essence, from which about two-thirds has been removed. It is also interesting to note that it compares very favourably with pure dried milk.

		Flesh Formers.	
Dr. Johnson's Analysis	Cocoa Nibs	23	Out of every 100 parts.
	Dried Milk	35	
COCOA ESSENCE		34¾	
Best French Chocolates		11	

Mr. Faussett, M.B., F.R.C.S.I., in a paper

read before the Surgical Society of Ireland, May, 1877, draws the attention of the Faculty to this subject, in connection with the feeding of infants :—

"Without presuming to pass any judgment on the many artificial substitutes which on alleged chemical and scientific principles have from time to time been pressed forward under the notice of the profession and the public to take the place of mother's milk, I beg to call attention to a very cheap and simple article which is always easily procurable—viz., Cocoa, and which ***when pure and deprived of an excess of fatty matter***, may safely be relied on, as Cocoa in the natural state abounds in a number of valuable nutritious principles; in fact, in every material necessary for the growth, the development, and sustenance of the body." After giving some remarkable cases of children being restored from "the last stage of extreme exhaustion" by its use, and "continued through the whole period of infancy" with the effect of their becoming fine, healthy children, he concludes by saying:—

"I beg, therefore, respectfully to commend Cocoa, as an article of infant's food, to the notice of my professional brethren, especially those who, holding office under the Poor-laws, have such large and extensive opportunities of testing its value."

For athletes, and all who study the development of the muscular tissues of the body, its use cannot be set aside. Professor Cavill, in his celebrated swim across the English Channel, and from Southampton to Portsmouth, considered it to be the most concentrated and sustaining food he could use for that trying test of his staying power; several other instances could be given, of the same character.

John Muter, Ph. D., F.C.S., in an article on Prepared Cocoa, says:—

"The only objection which can and does exist to its use in a state of purity is the excessive proportion of fat, which renders it too rich for most digestions, and gives unfortunately a colourable excuse for its admixture with starch. There are two classes of prepared Cocoa: (1) That in which the reduction of the fat is secured by adding starch and sugar; and (2) That where the fat is partially removed and the remainder of the bean is served to the public unmixed.

"On looking at the composition of Cocoa, the great fallacy of countenancing the addition of starch is at once apparent. The only possible excuse is the dilution of the fat, but then, at the same time, the nutritious gluten and stimulating theobromine are equally reduced in value. On the other hand, given the removal of a portion of the fat, the other constituents are not only kept intact but positively concentrated in a high degree."

Fine Cocoa, carefully prepared and combined with sugar, is probably the most delicious and delicate of all confections, and if free from the husk or shell, which is often used in the lower qualities of chocolate, is certainly one of the most nutritious articles of food.

The important question of adulteration may

F

Cocoa adulterated with common Arrowroot, containing Potato Starch, as seen by 1-5th inch power, and A eye-piece.

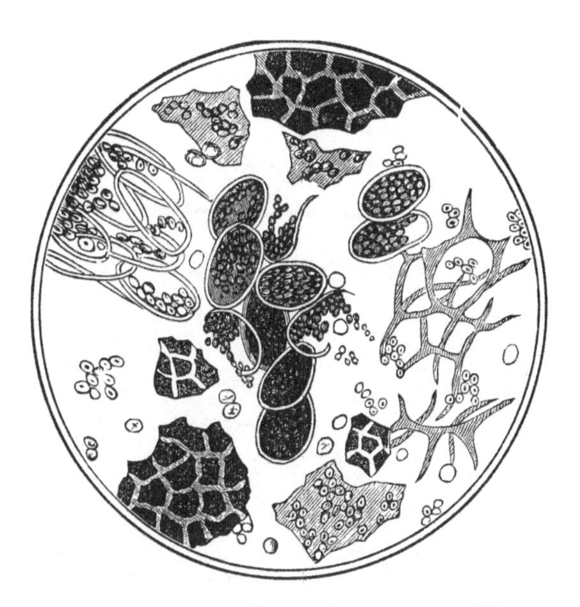

Pure decorticated Cocoa, with 1-5th inch power, and A eye-piece.
Starch cells—inner membrane—portions of embryo.

be divided under two heads—viz., those additions that are fraudulent but not injurious, and those additions that are both fraudulent and injurious to health.

Of this latter class of adulteration, Dr. Hassall's book on "Food and its Adulterations," written many years ago, well disposed of them, as it resulted in public opinion being awakened to such frauds, and in a searching investigation on the part of the Government. The Adulteration Act of Parliament (1873) made it necessary for the manufacturer to state on every packet that the article is sold as a mixture, and that all additions are in no way injurious to health.

We are glad to believe that Venetian red, umber, peroxide of iron, and even brick-dust, are adulterations of the past.

We have, therefore, to deal with the more difficult definitions of adulterations that are not necessarily injurious to health, but that reduce the value of Cocoa as food. We have already shown that Cocoa is rich in its nitrogenous elements, and therefore of such inestimable importance as a builder-up and strengthener of the human frame that we need be very jealous of all devices used by manufacturers to reduce its value. There are a certain class of additions to Cocoa that can only come under the head of fraudulent adulterations when they are mixed in extravagant quantities; we allude more particularly to farinaceous substances, such as arrowroot, sago, potato-starch, &c.

There can be no objection to such additions so long as they are stated, and the mixture not sold as Cocoa. It is hardly fair, however, to term such articles Chocolate, or Chocolate powder (certainly not Cocoa) when the proportion of Cocoa does not amount to one-tenth part of the whole.

We have samples of such articles before us that have been palmed on to the public as "Soluble Cocoa." We cannot admit under any circumstance that the addition of starch improves the quality. It may be that some prefer their Cocoa thick, but while the addition of farinaceous substances are made to absorb the excess of butter and to make it "soluble," it must not be forgotten that it takes the form of an emulsion, and therefore cannot be so wholesome as food.

We now come to the still more difficult and subtle question of the addition of alkali in the preparation, of pure Cocoa partly deprived of its natural butter. The reason for this is quite apparent to the initiated.

The addition of soda, potash, magnesia, or ammonia, either when the Cocoa is being roasted, or after that process, so acts upon the Cocoa that it deepens the colour, and gives an apparent strength when prepared for drinking, while it saponifies the butter still remaining in the Cocoa, thus holding the Cocoa longer in suspension. (Cocoa can never be chemically or actually soluble.)

An eminent English physician (Dr. Crespi) has lately written upon this question in a communication to the October, 1890, number of a leading American publication,

"Hygiene," and gives us the result of his investigations, from which we make the following extract:—

"Unfortunately we have of late years seen the country flooded with foreign Cocoas, pure in great measure—that is, innocent of starch and sugar—but contaminated with an admixture of alkali. The exact percentage of these additions and the steps in the process are not, however, perfectly clear. The object of this adulteration is this: Cocoa does not give an infusion or decoction, but mixed with water is practically a soup; it is suspended, not dissolved. Now the addition of an alkali gives rise to a soap in plain English, much as when common soap—a compound of oil and alkalies—is mixed with water; but this alkalised Cocoa has an appearance of strength which it does not possess, and the ignorant consumer hastily assumes that he is getting far more for his money and being supplied with a much better article, so that he cheerfully pays a higher price for his medicated beverage. But we are not so much concerned with the actual injury done by the adulteration of Cocoa with alkalies as with the principle. The recent great improvements in the preparation of Cocoa, as we have said above, by removing the superabundant oil, have so much increased the digestibility of this nutritious beverage that the last excuse for the addition of alkalies and starch is gone, and the presence of the former, besides being deleterious to some constitutions, cannot answer any purpose except giving an appearance of fictitious strength."

It may be as well also to add the opinion of Dr. Sidney Ringer, Professor of Medicine at the University College, London, and Physician to the College Hospital; perhaps the greatest English authority on the action of drugs.

He states, in his "Handbook of Therapeutics," that "the sustained adminstration of alkalies and their carbonates renders the blood poorer in solid and in red corpuscles and impairs the nutrition of the body." Of ammonia, carbonate of ammonia, and spirits of ammonia he says: "These preparations have many properties in common with the alkaline potash and soda group. They possess a strong alkaline reaction, are freely soluble in water, have a high diffusion power, and dissolve the animal textures."

One word with respect to the names "Cocoa" and "Chocolate." Cacao, or Cocoa, is the commercial name by which the plant and the fruit is known all over the world; therefore it directly applies to pure preparation of the fruit. When mixed with sugar or any other substance it cannot logically retain its original name, and the public have accepted this reasoning by asking for "Chocolate" when they require sweetened Cocoa. It would save much confusion if this rule was accepted and acted upon by manufacturers, or even enforced by Act of Parliament.

Much more might be written on the subject of adulteration, and we trust it will still claim the careful attention of those interested in the supply of pure and wholesome food.

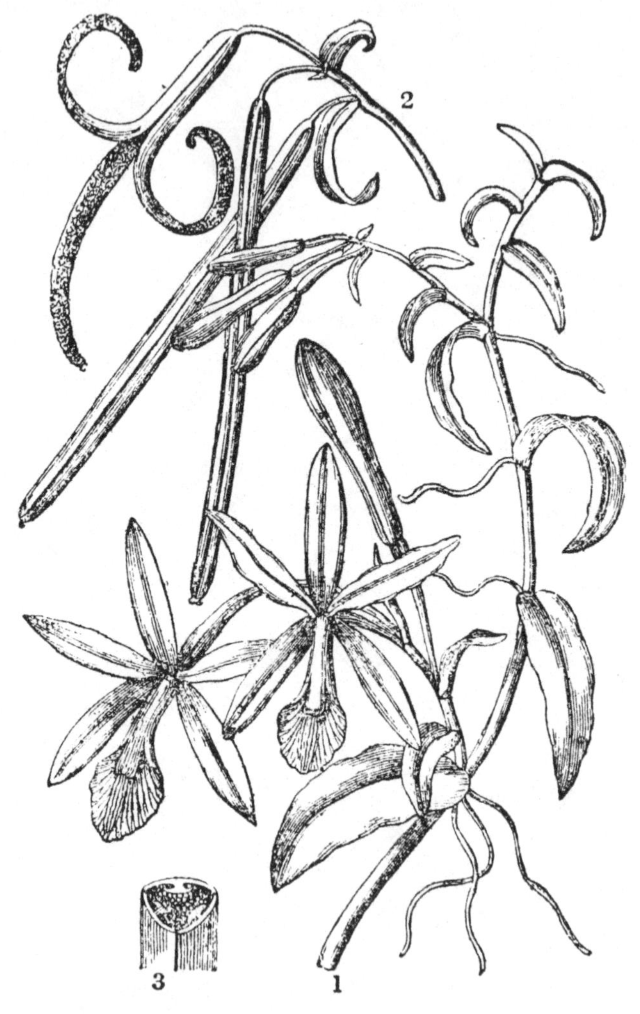

Vanilla Aromatica.—1, branch with flowers. 2, branch with fruit. 3, section of fruit showing the three placentæ and indefinite seeds.

VANILLA AROMATICA.

—:o:—

VANILLA is so intimately connected with the manufacture of chocolate, that the subject is hardly complete without some allusion to it in this place.

The name is derived from Vaynilla, which in Spanish, signifies a little knife, or scissor case, in reference to the shape of the pod.

Its natural habitat is the mountainous parts of Brazil. It is now cultivated in large quantities in other parts of the tropical world, the principal sources of supply coming from the French Colony of Reunion, Mauritius, and Seychelles, Bourbon, the West Indies, Java, Japan, and Madagascar, and within the past few years Ceylon and India all contribute to our supply. Mexico, from whence the principal supply used to flow, has almost ceased to grow Vanilla, owing to the comparatively low prices now obtained for it.

There are many species of this lovely and fragrant plant which belongs to the genera of Orchidaceous plants; but they differ in some respects to orchids generally, as the stem will grow to the height of from twenty to thirty feet. In climbing up the trees, the roots which they put forth as holdfasts are capable of absorbing nutriment for the plant when other modes of supply are cut off.

The places chosen for a plantation are shaded and warm ravines in high damp forests, taking for protectors trees with a soft bark, into which the roots may easily penetrate.

It is interesting, in connection with our subject, to hear from one who has cultivated the plant that he has seen it growing freely round the stem of the Cocoa tree like a hop.

Vanilla Aromatica and *V. Planifolia* are

the species from which the best kind of Vanilla is grown for commerce. The leaves are thick and fleshy, as are also the flowers, which are of a whitish-green colour.

The sweet perfume of its fruit is perceptible at a great distance, and attracts numbers of brilliantly coloured birds that dispute for the seeds when the fruit opens.

The Chica Vanilla of Panama is yielded by another orchid, a species of sobralia. The expressed juice of *V. claviculata*, a native of mountainous woods in the West Indies, is applied to recent wounds, and is hence called by the French in St. Domingo *Liane a blessures*. There is a species known as *zizpic* in Yucatan, which is a great ornament of the *cenotes*, or subterranean water caverns of the country. These singular caverns are sometimes entirely subterranean, and are then, of course, without vegetation; frequently, however, they are more or less open at the top, when they are often of surpassing beauty, on account of the luxuriant development of vegetable life which they contain. To these *cenotes* the few ferns of Yucatan are almost confined, and it is here that this Vanilla attains perfection. The pods are occasionally taken to market at Valladolid, where they may be bought at an almost nominal price.

Vanilla was not known in Europe until after the discovery of America, and little or nothing was known of the plant that produced it till 1703, when it was described by Plumier.

De Menonville, who travelled to Guaxaca in 1777, thus describes his discovery of Vanilla in that district. After various hindrances and disappointments, he says:— "At length an Indian, with a hoe in his hand, made his appearance. 'Brother,' said I, holding out a dollar, 'show me some Vanilla and this is yours.' He coolly bade me follow him, and advancing a few steps through the underwood into a thicket, in which were a number of trees, he immediately climbed up one, threw down to me two pods of Vanilla perfectly ripe, and pointed out to me a branch on which several others were hanging yet green, together with two faded flowers. The form of the leaves, the fruit, the peculiar smell of the plant—everything convinced me it was the real Vanilla in everything corresponding with such as I had seen at Vera Cruz. All the trees of this little copse were covered with it. I saw a quantity of green fruit, but collected no more than six specimens of these, and four large pods which were ripe. I caused the Indian afterwards to part from the root some of the scions which had sprung up. These I tied well together, wrapping up the whole in the leaves of an arum, which at their base are 3 feet wide. After thus packing a faggot, which weighed upwards of thirty pounds, I placed it in my large sack, which I fastened on my horse. I was so well satisfied with my Indian that, besides what I promised him, I gave him two reals in addition. For his part, unwilling to be outdone in generosity, he ran to his hut and brought me three other pods of vanilla."

The Duke of Marlborough introduced the plant into this country in 1800, from whence it made its way to the Continent where it is grown, and made to bear abundance of fruit.

At Liége it is grown on a small scale, and a plant cultivated at Paris in 1840 attained the height of three yards, and yielded 117 pods, which ripened in twelve months. Fine examples may be seen in the tropical and economic houses at Kew. Mr. Ewing and Mr. E. Bennett grew the Vanilla with considerable success at Csberton; the latter gathered no less than 300 ripe pods off a single plant in one season. He considers a temperature of from 50 to 70 degrees to be most suitable for it. He found it necessary to effect fertilization by artificial means, the stigma being prevented from receiving the pollen of its own flower by the interposition of an organ called the *retinaculum.*

This process, which in these climates is obliged to be performed artificially, is done naturally by insects in countries where the plants were originally found in their wild state.

The fruit of the plant is a long bean or pod, growing from four to twelve inches in length, and containing an immense number of small black granules, surrounded by a thick balsamous substance, which contains a peculiar volatile oil, and a considerable quantity of benzoic acid.

These give the delicious flavour and aroma, which prove so powerful and penetrating that a few ounces will flavour one hundred-weight of Chocolate.

Vanilla acts as a slight stimulant in the system, and the fact of its possessing benzoic acid is one that should not be lost sight of. It is said to be used by the Spanish Physicians in America as an antidote to poisons, and to the bite of venomous creatures.

The harvest in Mauritius begins in July, and as late as December in South America, the fruit being gathered when yellow, and after slight fermentation they are laid in the sun to dry; when about half dried, the pods are rubbed with mahogany nut oil, and after again being exposed to the sun are oiled a second time.

In another mode of preparation, the fruit for the market is allowed to dry until the pods have lost their yellowish green colour. Straw mats, covered with woollen blankets, are then laid on the ground, and when these are warm, the fruits are spread on them and exposed to the sun. After a time they are wrapped in blankets, and placed in boxes covered with cloth, after which they are again exposed. In about twelve hours, the fruits should become a dark coffee colour, but if they do not the process is repeated.

About fifty pods are then tied tightly together in a bundle, at each end, and once round the centre, with a species of grass, and packed in tins, which are hermetically sealed for export.

After some months, the pods become encrusted with an effloresence of white crystals, which possess properties similar to those in benzoic acid; they form very beautiful objects when seen through a microscope with polarized light.

There are various plans used for the drying and preparing the pods. All the care and success during the early cultivation may easily be lost through want of knowledge or watchfulness in these matters. We therefore refer the reader to extracts from a paper addressed to the Colonial Secretary, by the Assistant Director of the Royal Botanical Gardens, Mauritius, which gives a most exhaustive account of the planting, watering, ripening, harvesting, curing, and preparation for the market. *(See Appendix.)*

The South American wild Vanilla, or Vanillon, as it is generally called, is the variety *V. Sylvestris*, and is a much larger and coarser variety than those grown for commerce. In some parts of Brazil the women entwine this species of Vanilla bean in their hair; a weakness for fragrance that is common all the world over, although in this case it hardly fits in with our ideas of good taste.

A great variety of spices are used for the flavoring of Chocolate, but Vanilla still holds the palm. In Brooke's translation from the French, 1730, he tells us :—"The Spaniards try'd to make it (Cocoa) more agreeable by the addition of sugar, some Oriental spices and things that grow there, which it will be needless to mention—there is none continued down to us but Vanilla; in like manner that Cinnamon is the only spice which has had general approbation, and remains in the composition of Chocolate."

The old French writer is very strong in his denunciation of Vanilla, as he goes on to say :—"Whereas Chocolate season'd with Vanilla, and other hot and biting ingredients, cannot but be pernicious, especially in summer, to young people, and to dry constitutions."

The French still name Chocolate prepared without Vanilla "Chocolat Santè," but we have fortunately lived down this prejudice, and Vanilla reigns supreme among all ingredients used to flavour Chocolate.

It has already been intimated, with respect to the growth of Vanilla in Mexico, that prices have fallen so considerably that it barely pays to grow the crop in that country. Many years ago prices ranged as high as 120/- per pound on the market; the highest price now does not realise one fourth that sum. This is due to the introduction of a beautiful crystalline substance called Vanilline, which coincides in almost every particular with the active principles contained in Vanilla.

The distinctive aroma is so nearly the same that when incorporated with Chocolate it is difficult to distinguish with which it has been flavored. It would, however, be going too far to assert that it is equal in aromatic and fruity flavour to Vanilla.

One ounce of Vanilline crystals are about equal in flavoring power to one pound of good Vanilla beans, the market value is therefore about in the proportion of sixteen to one.

Vanilline, of which the chemical formula is $C^8 H^8 O^3$, is prepared from Coniferine, which is to be found in considerable quantities in the plants of the numerous family of the Conifers.

Coniferine was discovered by Hartig in 1861 in the sap of the *Larix Europea*; later on, its presence was recognised in all species of pines and firs.

In 1874 Messrs. Haarman and Tiemann showed that the Coniferine, under the influence of oxydizing agents properly chosen, could be made to produce Vanilline, which is no other than the aromatic principle of the Vanilla pod. This discovery has given a certain industrial importance to Coniferine, and it has already been collected by hundreds of kilogrammes in the forests of North Germany.

The first consignments of Vanilline were between 1874 and 1878 and were prepared by a patent process, that is to say by the oxidization of the Conferine. This is only to be found in the descending sap of the pines, so in the spring incisions were made at the base of the trees and the sap which flows from them was collected; it was then filtered and exposed to the air, when it soon became solid. In this state it constitutes the Conferine, which could be preserved indefinitely and made use of when required for its transformation into Vanilline.

The more modern and advantageous plan is to fell the tree, cut off the branches, and strip them of their bark. The sap is then collected by scraping the trunk with a sharp instrument—an iron scraper or a knife—and the liquid, as it oozes out is absorbed by a coarse sponge, and then squeezed into a tin bucket. If too long a time elapses between these two processes the evaporation is rapid enough to solidify the juice and then it cannot be collected. The sap presents the appearance of a white milky opaque liquid, and in its normal state contains a particular sugar, albumen, and Conferine. In order to prevent fermentation it ought after five or six hours at the most to be boiled in a furnace from ten to fifteen minutes, so as to congeal the albumen it contains. The boiling liquid is filtered through a coarse flannel or baize bag, and the filtered liquid is then evaporated to the fifth part of the original quantity. It is then allowed to cool gently in a shady place for one night; it then deposits very small white crystals of Vanilline. In order to collect them the liquid is thrown upon a linen cloth, and when the crystals are sufficiently drained they are pressed in order to squeeze out the brown syrup which colours them and prevents their drying.

The Vanilline obtained by this method is identical with the crystals already described as forming on the Vanilla pod, the chemical constitution of the one being identical with the others, as also their physical properties.

APPENDIX.

Extracts from a paper addressed to the Colonial Secretary by Mr. N. Cantley, Assistant Director of the Royal Botanical Gardens, Mauritius, and supplied by Messrs. Brookes and Green, Brokers, Mincing Lane, London.

July 23, 1874.

SIR,

I have the honor to lay before you the following details respecting the cultivation of the Vanilla Plant (*V. Planifolia*) as practised by the principal growers in this Colony, viz.:—

PREPARING THE GROUND FOR PLANTING.

The plant will grow tolerably well in any porous soil, still it has been found by practical growers that a composition consisting of equal parts of well decomposed leaves, loam, sand, and charcoal, is best suited to the wants of the plant, and when this can be obtained trenches should be made the entire length of the ground intended for the plantation, 18 inches wide, 2 feet deep, and 8 feet apart, and filled previous to planting with the composition just alluded to. Some growers put only a small quantity of the composition into the trenches the first season, or sufficient to give the plants a start, adding the remainder year by year, by way of surface dressing, but this is objectionable in countries subjected to heavy periodical rains, as the trenches often stand full of water during such rains, greatly to the injury of the plants; it is, therefore, safer to fill the trenches the first year, or when the plants are planted.

SHADE.

The Vanilla, like the rest of the orchidaceal, delight in shade, a fact which at once suggests that it ought to be planted among trees sufficiently large to screen it from the direct rays of the sun, but where such trees are not available young trees must be planted, and now arises the question, what will grow quickest in order that the Vanilla may be planted as soon as possible? The plants most commonly used for this purpose in Mauritius are the *Lilas de l'Tude* of the Creoles, *Melia Azadarech*, and *Tecoma Lencoxylon*, and are planted 8 feet apart, as permanent plants, but when these are only a few inches high when planted it is evident that a period of at least three years must elapse before any considerable amount of shade can be expected from them, and this would delay the planting of the Vanilla an equal length of time had not the planters access to another plant, the *Pignon de l Tude* of the Creoles (*Tatropha Curcus*), which is of extremely rapid growth, but of no permanent nature; it is easily increased by cuttings of the branches, which are generally cut in lengths of 2 feet, and planted, 18 inches apart, in line with the *Lila* and *Tecoma* plants previously mentioned, and as they will very soon produce leaves, some growers plant the Vanilla at the same time, and train it under the shade of *Pignon de l' Tude.*

It is better to wait until the *Pignon de l' Tude* be sufficiently strong to allow of the trellis work being erected. Where dead palm leaves are plentiful some growers screen the whole plantation at the first outset, and plant the Vanilla at once, but it is not often that leaves are to be had in sufficient quantity to allow of this being done, and, again, it is ten to one if the first strong wind does not destroy the whole construction.

PLANTING.

The usual method of planting Vanilla is by cuttings of the stems of strong healthy plants, and if cut in lengths of 3 feet, they will generally produce fruit 18 months after planting. Plantations are generally made during the months of October and November, in Mauritius, or at the commencement of the hot season, when the sap, after a season of comparative rest, is being stimulated by the increasing heat into renewed action. As the rapidity of growth greatly depends on the number of roots, care must be taken that three joints (nodes) of the cutting be placed on the ground, in an oblique direction, and from these joints a plentiful supply of roots will be given out, which must be treated as hereafter stated.

TRAINING, OR ARTIFICIAL SUPPORT.

The rapidity with which the Vanilla plant grows when its roots have free action in a suitable soil is somewhat amazing, it is therefore no question that, if some system of artificial training be not resorted to, it must very soon outgrow the plants which are intended to shade it. Few trees can keep pace with the Vanilla plant, and even were it otherwise, it would not be advisable to let the plant grow straight up, because they would very soon get out of reach in this position, so that a ladder would have to be used in gathering the pods, fertilizing, &c., but, on the other hand, the plants ought to be allowed to run straight up until they arrive at about 6 feet in height, as it is well known that the more perpendicular the plant be the quicker will it grow. But to prevent this taking place poles of about 7 feet in length are driven into the ground, perpendicularly, about 8 feet apart, and when well firmed horizontal bars are attached to them, one foot apart. As soon as the Vanilla has reached the top of this con struction it is trained along the horizontal bars and thus prevented from getting too much sun.

WATERING.

The quantity of water required is greatly modified by the state of the weather and kind of soil in which the plants are growing, but if planted in the composition previously recommended, a good watering twice a week in the hot season, and once a week in the cold season, will be generally found sufficient, but should never be given until the ground be found thoroughly dry, as nothing will kill the plants sooner than stagnant water at their roots.

MANURING AND SURFACE DRESSING.

Of the various manures applied to the Vanilla, nothing has been found to suit the plants better than vegetable mould, especially when mixed with a little charcoal or wood ashes, and the plants will be greatly benefited by a surface dressing of this manure in the second year of the plantation, at which date, should any of the roots be found to have penetrated the natural soil lying between the trenches it should be turned over with a digging fork, adding at the same time a liberal quantity of manure, but should the roots be found not to have reached such soil, it will be sufficient to remove the surface to a depth of about 2 inches, and 2 feet in width on each side of the Vanilla stems, and replace with manure.

FERTILIZATION.

Self fertilization in the Vanilla, especially in *V. Planifolia*, is of rare occurrence to any useful extent owing to the intervention of the reticulum, which is a portion of the inner face of the style, which makes artificial fertilization an indispensable process, and which simply consists in removing the pollen from the anthers of the flower and applying it to the stigma with a small camel hair brush, this is best done about mid-day during bright sunny weather.

A brush is used in England, but the anther itself may be applied. Some growers look after the fertilization in the cool of the morning, if this be preferred a dry morning should be chosen.

DURATION OF THE PERIOD OF RIPENING.

The duration of the period of ripening from time of fertilization is generally about nine months, when the pods become of a yellowish green colour, and the valves show a tendency to open (dehise) at the lower extremity, which is a sure sign of ripeness.

HARVESTING THE PODS.

Most of the pods are harvested in Mauritius during the month of July, but as the plants come into flower very irregularly, all the pods cannot be harvested at the same time, so that those found unripe at the first gathering must be waited for until they show the usual signs of ripeness, for if gathered sooner they will shrivel during the process of drying, and fetch but a small price in the market, should any of the pods, however, be left on the plant until the valves have opened (if more than half an inch they are useless) they should be at once removed, the opening part tied up, and plunged for a moment in boiling water, and afterwards put through the usual process of drying with the other pods.

DRYING AND PREPARING THE PODS.

The drying and preparing of the pods is undoubtedly a most difficult item in Vanilla cultivation, and if not thoroughly understood all previous labour with the plant will be lost.

The different methods of preparation, as practised in Mauritius, are more or less as follows. A large oven (similar to that in which ordinary bread is baked) with two valves on the top, and a thermometer fixed in the iron-work in front, is heated by the application of fire beneath until the thermometer indicates a heat of 40 degrees centigrade, or 104 degrees fahrenheit. The fire is then withdrawn, and pods, in quantity of about 60 or 70 lbs. together, are well wrapped in Banana leaves, which are again covered with a woollen cloth, and placed in the oven, where they are allowed to remain until the mercury of the thermometer falls to about 90 degrees fahrenheit, after which they are removed and placed in a wooden box to sweat, and cooled gradually; when nearly cold they are exposed to the sun on boards covered with blankets, which when warmed by the sun, the pods are spread on them.

Some growers, instead of letting the pod be exposed to the direct rays of the sun, cover them with a black woollen cloth, which by absorbing the great amount of heat from the sun, as well as moisture from the pod, modifies the process of drying, and imparts to the fruits a more superior flavor and color than they otherwise would have; after being thus exposed for two or three days, they ought to be of a dark brown, or coffee colour, they are next laid on perforated shelves in an airy room, where they are allowed to remain for one month, or until dry, when they will be found to have shrunk to one fourth of their natural size, and of a blackish hue with a somewhat silvery appearance.

PREPARING THE PODS FOR MARKET.

In preparing the pods for market those of a size are carefully selected, and tied in bundles of fifties, good marketable pods are about 8 inches long, and 50 of those will generally weigh three-quarters of a pound, the smaller pods, half-a-pound. After being thus arranged they are packed in tins, about 12 or 16 lbs. each, and sent to the London market.

Printed in the United States
By Bookmasters